A LOOK AT LIFE SCIENCE

WHAT ARE FUNGI?

BY KATE MIKOLEY

Gareth Stevens PUBLISHING

CRASHCOURSE

Please visit our website, www.garethstevens.com. For a free color catalog of all our high-quality books, call toll free 1-800-542-2595 or fax 1-877-542-2596.

Cataloging-in-Publication Data
Names: Mikoley, Kate.
Title: What are fungi? / Kate Mikoley.
Description: New York : Gareth Stevens Publishing, 2020. | Series: A look at life science | Includes glossary and index.
Identifiers: ISBN 9781538248553 (pbk.) | ISBN 9781538248577 (library bound) | ISBN 9781538248560 (6 pack)
Subjects: LCSH: Fungi--Juvenile literature.
Classification: LCC QK603.5 M63 2020 | DDC 579.5--dc23

First Edition

Published in 2020 by
Gareth Stevens Publishing
111 East 14th Street, Suite 349
New York, NY 10003

Designer: Reann Nye
Editor: Kate Mikoley

Photo credits: Series art MriMan/Shutterstock.com; cover, p. 1 Regina Nogova/Shutterstock.com; p. 5 FotograFFF/Shutterstock.com; p. 7 Andreas Zerndl/Shutterstock.com; p. 9 Poliorketes/Shutterstock.com; p. 11 Isabel Eve/Shutterstock.com; p. 13 ROMAN KALISHCHUK/Shutterstock.com; p. 15 dugdax/Shutterstock.com; p. 17 Olena Belevantseva/Shutterstock.com; p. 19 SpeedKingz/Shutterstock.com; p. 21 mahey/Shutterstock.com; p. 23 Peleg Elkalay/Shutterstock.com; p. 25 O_Solara/Shutterstock.com; p. 27 Kallayanee Naloka/Shutterstock.com; p. 29 MyImages - Micha/Shutterstock.com; p. 30 (left) Svitlyk/Shutterstock.com; p. 30 (middle) kwanchai.c/Shutterstock.com; p. 30 (right) Chongsiri Chaitongngam/Shutterstock.com.

Printed in the United States of America

CONTENTS

Words in the glossary appear in **bold** type the first time they are used in the text.

FUN FUNGI

If you see a living thing growing in the ground, you may think it must be a plant. However, that's not always the case. Fungi are **organisms** that commonly look like plants, but are actually quite different from them.

MAKE THE GRADE

The word "fungi" is plural. This means we use it when we're talking about more than one fungus. "Fungus" is the **singular** form of the word.

SORTING THINGS OUT

Scientists sort living things into groups based on things they have in common. This is called classifying. Fungi were once classified in the same group as plants. By the later part of the 1900s, scientists realized fungi belonged in their own group, or kingdom.

MAKE THE GRADE

Fungus is a Latin word meaning "mushroom." A mushroom is a kind of fungus

MAKING MORE FUNGI

Most fungi form spores. These are sort of like seeds and are commonly made up of one cell. Fungi **release** many spores. When a spore leaves a fungus and lands in a place where it can live, it grows into a new fungus.

MAKE THE GRADE

Most fungi live on land, commonly in damp areas. However, some species, or kinds, live in water.

FEEDING TIME

Unlike plants, fungi cannot make their own food. Instead, they get their food from things around them. Fungi give off kinds of matter called enzymes. The enzymes break down the food. Then, the food is **absorbed** through parts of the fungi.

MAKE THE GRADE

Many fungi get **nutrients** from plants or animals that have died, such as trees that have fallen.

When fungi get their food from dead matter, they're helping the matter to decompose, or break down. When a dead organism is decomposed, it's broken down into matter that living things in the **environment** can use.

MAKE THE GRADE

Fungi that decompose dead plants and animals help clear away matter that would otherwise cover large parts of the forest.

PARASITE PESTS

Some kinds of fungi get their nutrients from other living things. These are called parasites. Commonly, these fungi feed on plants. This can hurt or even kill the plants. At times, parasitic fungi have ruined crops and food **sources**.

MAKE THE GRADE

In the United States, parasitic fungi have destroyed many chestnut and elm trees.

CHESTNUT TREE

Some parasitic fungi can be harmful to animals. When fungi **infect** plants it's not always easy to see. If animals eat the plant, they could become sick or die. In Europe during the **Middle Ages**, many people became sick from eating fungi-infected rye bread.

MAKE THE GRADE

Today, methods can be used to remove infected parts from plants so people and other animals won't get sick from them.

IN HIDING

The part of the fungus you can commonly see is called the fruiting body. Often, much more of the fungus grows underground. In Oregon, there's a fungus that covers more than 2,000 acres (809 ha), but most of it can't be seen because it's underground!

MAKE THE GRADE

Some fungi are so small they can only be seen with a special tool called a microscope.

AMAZING MUSHROOMS

Many fungi have a fruiting body shaped sort of like an umbrella. These are called mushrooms. It's common for people to eat some kinds of mushrooms. Other kinds are poisonous, or toxic, and could be very harmful if they are eaten.

MAKE THE GRADE

You should never eat a mushroom you find growing in the wild, as it could be poisonous. Poisonous mushrooms are often called toadstools.

BAKING WITH FUNGI

Yeast is a kind of fungus made up of one cell. It can be found in soil and on parts of plants all over the world. Some kinds are used in making foods such as bread. It's what helps make bread rise.

MAKE THE GRADE

Yeast used in baking feeds on sugar and then makes **carbon dioxide**, which causes the food to rise.

WHAT'S MOLD?

If you've ever had to throw away a piece of bread or fruit because it was covered in a fuzzy greenish matter, you've seen fungi. This kind of fungus, called mold, is often a sign that food is too old to eat.

MAKE THE GRADE

While some molds aren't safe to eat, others are used to make some cheeses!

Some kinds of mold are able to kill harmful bacteria. One kind, called *Penicillium*, is used to make a **medicine** called penicillin. This is used as an antibiotic, or a drug that can kill kinds of bad bacteria.

MAKE THE GRADE

Penicillin is often used to treat illnesses such as throat infections.

MORE TO FIND?

Fungi can come in all sorts of shapes and sizes and live in many kinds of **habitats**. There are about 144,000 species of fungi known about in the world. However, scientists think there may be millions more we don't yet know about!

MAKE THE GRADE

Leafcutter ants bring pieces of leaves back to their nests. They then eat a fungus that grows on the leaves!

FUNGI: GOOD OR BAD?

HELPFUL FUNGI

- help make foods such as bread and cheese
- make medicine
- decompose dead matter
- when safe to eat, can be tasty and provide nutrients

HARMFUL FUNGI

- can hurt plants and animals
- can make people sick
- can grow on food, making it unsafe to eat

GLOSSARY

absorb: to take in

carbon dioxide: a gas breathed out by animals

environment: the natural world in which a plant or animal lives

habitat: the natural place where an animal or plant lives

infect: to spread something harmful

medicine: a drug taken to make a sick person well

Middle Ages: a time in European history from about 500 to 1500

nutrient: something a living thing needs to grow and stay alive

organism: a living thing

release: to set something free

singular: the form of a word used when talking about only one thing

source: something that offers a supply of a particular thing

FOR MORE INFORMATION

BOOKS

Axelrod-Contrada, Joan. *Mini Mind Controllers: Fungi, Bacteria, and Other Tiny Zombie Makers*. North Mankato, MN: Capstone Press, 2017.

Twiddy, Robin. *Mushrooms and Fungi*. New York, NY: Kidhaven Publishing, 2019.

WEBSITES

Biology for Kids: Fungi
www.ducksters.com/science/biology/fungi.php
Read more about fungi and take a quiz on this website.

Mushrooms
www.dkfindout.com/us/animals-and-nature/mushrooms/
Learn about many different kinds of mushrooms here.

Publisher's note to educators and parents: Our editors have carefully reviewed these websites to ensure that they are suitable for students. Many websites change frequently, however, and we cannot guarantee that a site's future contents will continue to meet our high standards of quality and educational value. Be advised that students should be closely supervised whenever they access the internet.

INDEX